Faucher der Saint-Maurice

The Province of Quebec and Canada at the Third International

Geographical Congress

at Venice

Faucher der Saint-Maurice

The Province of Quebec and Canada at the Third International Geographical Congress
at Venice

ISBN/EAN: 9783741123405

Manufactured in Europe, USA, Canada, Australia, Japa

Cover: Foto ©Klaus-Uwe Gerhardt /pixelio.de

Manufactured and distributed by brebook publishing software
(www.brebook.com)

Faucher der Saint-Maurice

The Province of Quebec and Canada at the Third International Geographical Congress

THE PROVINCE OF QUEBEC AND CANADA

AT THE

THIRD INTERNATIONAL GEOGRAPHICAL

CONGRESS

AT

VENICE

SEPTEMBER 1881

LEVIS
PRINTED BY MERCIER & CO.

1882

To the Honorable E. T. PAQUET,

Provincial Secretary,

Quebec.

Sir,

On the first of July 1880, the prince of Teano sent a circular, in the name of the Geographical Society of Rome, informing the various Governments and Scientific Societies that the third International Geographical Congress would be held at Venice in September 1881.

The first meeting had taken place at Antwerp in 1871 and the second at Paris in 1875. His Excellency Count Viola, a devoted friend of this country, in which be resided for some time, caused an invitation to be sent to the Province of Quebec inviting it to take part in the exhibition.

In exchange for this courtesy, His Honor the Lieutenant-Governor in Council appointed Count Viola our representative in Venice, giving him as colleagues, with instructions to prepare the Canadian exhibit at Quebec, Mr. Eugène Taché, assistant Commissioner of Crown Lands, and Mr. Faucher de St. Maurice, Clerk of Private Bills in the Legislative Council.

The Congress was to be made up of the eight following scientific groups,

I. Geography, mathematics, geodesy, topography ;

II. Hydrography and maritime geography ;

III. Physical geography, meteorology, geology, botany and zoology ;

IV. Anthropological and philological geography ;

V. Historical geography and the history of geography ;

VI. Economical, commercial and statistical geography ;

VII. Methodology, teaching and diffusion of geography ;

VIII. Geographical explorations and travels.

We set to work and Mr. Taché sent to Venice the maps, books and specimens which the Government placed at our disposal.

The following is an extract from the official catalogue, respecting Canada, printed in Italy, under the direction of Count Viola ;

CANADA •

GOVERNMENT OF THE PROVINCE OF QUEBEC

*Works sent to the Exhibition through Messrs. E.
Taché, Assistant Commissioner of Crown Lands
and N. Faucher de St-Maurice, Clerk of Private Bills in the Legislative Council,*

CLASS III.

PHYSICAL GEOGRAPHY, METEOROLOGY, GEOLOGY, BOTANY, ZOOLOGY.

Geological survey of Canada (Sir William E. Logan, Director ; Alfred Selwyn, his successor ; Alex.
Murray, T. Sterry Hunt, E. Billings, members.)
1. Reports of progress from its commencement to
1863. Illustrated with woodcuts in the text ;
accompanied by an atlas of maps and sections.
Translated by order of the Government from
English into French by Professor Darey.
Montreal, 1864.

Geological Survey of Canada.
2. Reports of progress from 1876 to 1879. Published
by order of Parliament.

Geological Survey of Canada.
Sir W. E. Logan, Director; Alfred Selwyn, his
successor.
3. Report of progress from 1863 to 1866. Translated
from English into French and printed by order of
His Excellency the Governor General.
Printed by Desbarats 1866.
4. Report of progress from 1873 to 1876, published
by order of Parliament,
Montreal, Dawson, 1876.

Geological Survey (A. Selwyn, Director.)

5. Report of progress from 1866 to 1871, accompanied by geological and topographical maps. Translated from English into French under the direction of the commission.
   Printed by Taylor, 1873.

Abbé Laflamme, Professor of Laval University.
6. *Éléments de minéralogie et géologie.*
   (Elementary mineralogy and geology.)
   Quebec, Delisle, 1881.

A. P. Landry.
7. *Traité populaire d'agriculture théorique et pratique.*
   (Popular treatise on the theory and practice of agriculture.) A work for which a prize was given by the Council of Agriculture of the province of Quebec.
   Montreal, Canadian Printing Co. 1878.

Dr. G. Larocque.
8. *Manuel d'horticulture pratique et d'arboriculture fruitière.*
   (Manual of practical horticulture and the culture of fruit trees.)
   Lévis, Mercier, 1880.

Abbé L. Provencher, editor of *Le Naturaliste Canadien.*
9. *Le verger, le potager et le parterre dans la province de Québec.* A treatise on the cultivation of fruit, vegetables and flowers which are suited to the climate of Quebec.
   Quebec, Darveau 1881.
10. Geological Map of Canada. Large size.
11. Map of the Province of Quebec.
12. Mining Map of the Province of Quebec.
13. Geological Map of the Province of New-Brunswick.
14. List of Canadian fruit trees.
15. List of Canadian birds.
16. Specimens of various Canadian woods.

17. Specimens of phosphate from the county of Ottawa, province of Quebec.
17. *Guide du colon.*—Province of Quebec, 1880.

CLASS V.

HISTORICAL GEOGRAPHY, HISTORY OF GEOGRAPHY.

Abbé Casgrain.
19. *Une paroisse Canadienne au dix-septième siècle.* (A Canadian parish in the 17th century.) Quebec, Leger Brousseau, 1880.

CLASS VI.

ECONOMICAL, COMMERCIAL AND STATISTICAL GEOGRAPHY.

Honorable Mr. Chauveau (late minister of Public Instruction for the Province of Quebec.)
20. Public Instruction in Canada. A historical and statistical summary. Quebec, Côté & Co, 1876.
Arthur Buies.
21. *Le Saguenay et la vallée du lac St. Jean.* An historical, geographical, industrial and agricultural study.—Recent data and statistics; Picturesque descriptions with plates. Quebec, Côté & Co, 1880.
Paul de Cazes.
22. *Notes sur le Canada.*—History, population, products, commerce, navigation, railways, militia, &c., &c. Quebec, Darveau, 1880.
Abbé Ferland.
23. *Opuscules.*—New edition. Quebec, Côté & Co, 1877.
24. Census of Canada. Printed by Taylor 1873, 5 vols.
25. Report of the Minister of Agriculture for Canada 1880. Printed by order of Parliament.

26. General reports of the Commissioner of Agriculture and Public Works of the Province of Quebec, from 1868 to 1880. Printed by order of the Legislature.
27. Thirteenth annual report of the Minister of Marine and Fisheries 1880. Printed by order of Parliament.
28. Annual report of the Minister of Public Works for the year 1879-80. Printed by order of the House of Commons.
McLean, Roger & Co, 1881.
29. Annual report of the Minister of Railways and Canals for the year 1879-80, upon the work executed under his control. Printed by order of the House of Commons.
McLean, Roger & Co, 1881.
30. Statistics of the Canals during the navigation season of 1880. Supplement n. 1 to the report for the year ending 30th June 1880.
31. Report of the Post Master General for the fiscal year ending on the 30th June 1880. Printed by order of Parliament.
McLean, Roger & Co, 1880.
32. Public Accounts of Canada for the year ending 30th June 1880. Printed by order of Parliament.
McLean, Roger & Co, 1881,
33. Statement of the public accounts of the province of Quebec for the years 1878 to 1880. Printed by order of the Legislature.
34. Report of the Superintendent of Public Instruction of the Province of Quebec, from 1868 to 1880. Printed by order of the Legislature of Quebec.
35. Report of the Superintendent of Public Instruction of the province of Quebec for the year 1878-79. Printed by order of the Legislature.
36. Reports of the Commissioner of Crown Lands for the Province of Quebec from 1868 to 1880.
37. Tables of the Trade and Commerce of Canada for

the year ending 30th June 1880. Published by order of Parliament.
Printed by McLean, Roger & Co, 1881.
38. Reports, statements and statistics of the Internal Revenue of Canada for the year ending 30th June 1880.
Printed by McLean, Roger & Co, 1880.
39. Report of the Secretary of State for Canada for 1880. Printed by order of Parliament.
Printed by McLean, Roger & Co, 1881.
40. Report of the Auditor-General upon the credit account, &c., for the year ending 30th June 1880.
Printed by McLean, Roger & Co, 1880.
41. Report on the adulteration of food ; Supplement to Report No. 3 of the Minister of the Interior.
Printed by order of Parliament 1880.
42. Maps of the Province of Quebec shewing the railways.
43. Maps of the Province of Quebec shewing the lands conceded up to date.
44. Map of the projected Lower Laurentian Railway.

CLASS VII.

METHODOLOGY, TEACHING AND DIFFUSION OF GEOGRAPHICAL
KNOWLEDGE.

F. X. Toussaint (professor at the Laval Normal School).
45. An abridgement of modern Geography for the use of elementary schools. Approved by the Council of Public Instruction.
Quebec, Darveau, 1871.

F. X. Toussaint.
46. An abridgement of modern Geography approved by the Council of Public Instruction.
Quebec, Darveau, 1877.

F. X. Toussaint.
47. Abridged History of Canada, for the use of elementary schools in the Province of Quebec, published
by the Council of Public Instruction.
Quebec, Darveau, 1877.
I. G. Hodgins.
48. Geography and History of the British Colonies
(with plates).
Montreal, Lovell, 1866.
The Christian Brothers.
49. New illustrated Geography for the use of Christian
schools.
Montreal, Chapleau & Son.
L'abbé Holmes.
50. New abridgement of modern Geography, for the
use of youth, 8th edition, revised, corrected and
considerably increased by l'abbé L. O. Gauthier.
Montreal, Rolland & Son, 1877.
F. X. Garneau.
51. Abridged History of Canada, from its discovery to
the year 1840. For the use of schools. New edition.
Montreal, Beauchemin & Valois, 1876.
52. Elements of modern Geography, printed under the
direction of the Educational society of the district
of Quebec.
Montreal, Rolland & Son, 1877.
53. Plan of the City of Quebec.
54. Map of the Eastern Townships.
55. Map of the Lake St. John District.
56. Map to accompany the History of Canada.

CLASS VIII.

GEOGRAPHICAL EXPLORATIONS AND TRAVELS.

Faucher de Saint-Maurice.
57. *Promenades dans le golfe Saint-Laurent.* A portion

of the North Shore ; Egg Island ; Anticosti ; St.
Paul's Island ; Magdalen Island. 3rd edition.
Quebec, Darveau, 1880.

Faucher de Saint-Maurice.

58. *Promenades dans le golfe St. Laurent* : Nova Scotia,
Prince Edward's Island ; New Brunswick ; Baie
des Chaleurs, Gaspesia. 3rd edition.
Quebec, Darveau, 1880.

Faucher de Saint-Maurice.

59. *De Québec à Mexico.* Souvenirs of travels, garrisons,
battles and bivouacs. 4th edition.
Montreal, Duvernay, 1874, 2 vols.

Faucher de Saint-Maurice.

60. *De tribord à babord.* (*From port to starboard.*)
Three cruises in the Gulf of St. Lawrence, North
and South ; a portion of the North Shore ; ship-
wreck of Admiral Walker ; Anticosti ; Magdalen
Islands ; Nova Scotia ; New Brunswick ; Prince
Edward's Island ; Gaspesia,
Montreal, Duvernay, 1877.

CATALOGUE

of specimens of woods of Canadian Forests.

Botanical names.	English names.
1. Alnus incana........	Alder,
2. Pyrus Americana....	Ash Mountain.
3. Fraxinus juglandifolia.	Ash rim.
4. Fraxinus pubescens..	Ash swamp.
5. Fraxinus sambucifolia.	Ash black.
6. Fraxinus americana...	Ash white.
7. Populus grandidentata	Aspen Large-toothed.
8. Tilia americana......	Bass wood.
9. Fagus ferruginea.....	Beech.
10. Carpinus americana..	Beech blue.
11. Betula lenta vel nigra.	Birch black.
12. Betula excelsa.......	Birch curly.
13. Betula papyracea.....	Birch white.
14. Juglans cinerea......	Butternut smooth.
15. Platanus occidentalis..	Button wood.
16. Thuja occidentalis....	Cedar white.
17. Juniperus virginiana..	Cedar red.
18. Cerasus virginiana....	Cherry-Choke.
19. Cerasus serotina.....	Cherry (*Wild*) black.
20. Castanea americana..	Chestnut.
21. Aesculus hyppocastanum.............	Chestnut-Horse.
22. Cerasus pensylvania..	Cherry (*Wild*) red.
23. Populus monilifera...	Cotton wood (*necklace poplar.*)
24. Cornus florida..... .	Cornell (*Flowering Dog wood.*)
25. Ulmus americana....	Elm, grey or white.
26. Ulmus fulva vel rubra.	Elm red,
27. Ulmus fulva.........	Elm slippery.

Botanical names.	English names.
28. Ulmus racemosa.....	Elm soft.
29. Abies balsamea......	Fir Balsam.
30. Abies canadensis.....	Hemlock.
31. Carya tormentosa....	Hickory smooth bark.
32. Carya alba..........	Hickory rough bark.
33. Ostrya virginica......	Iron wood.
34. Acer saccharinum....	Maple hard.
35. Acer spicatum.......	Maple Mountain.
36. Acer rubrum........	Maple soft curly.
37. Acer dasycarpum....	Maple soft (*plane*.)
38. Acer pennsylvanicum.	Moose wood.
39. Quercus tinctoria.....	Oak black.
40. Quercus ambigua....	Oak Grey (*lake*.)
41. Quercus rubra.......	Oak red.
42. Quercus alba........	Oak white.
43. Quercus alba........	Oak white (*Ottawa*.)
44. Pinus resinosa.......	Pine red.
45. Pinus strobus........	Pine white.
46. Pinus mitis..........	Pine yellow.
47. Prunus americana....	Plum-wild, yellow.
48. Populus balsamifera..	Poplar Balsam or Balm of Gilead.
49. Populus tremuloides..	Poplar (*Common Aspen*.)
50. Sassafras officinalis...	Sassafras.
51. Pinus rupestris.......	Scrub pine.
52. Abies alba..........	Spruce white.
53. Abies nigra.........	Spruce black.
54. Rhus typhina........	Sumach.
55. Larix americana.....	Tamarac.
56. Cratœgus punctata...	Thorn Apple.
57. Cratœgus coccinea...	Thorn white.
58. Juglans nigra........	Walnut black.
59. Salix nigra..........	Willow black.
60. Liriodendron tulipifera	White wood.

In the month of July, I was obliged to go to France. I went to Venice, where I had the honor of taking part in the work of the Congress.

One of the rooms in the Royal Palace in the square of St. Mark was placed at the disposal of Canada. Count Viola had spared neither expense nor trouble to enable the province of Quebec to make a worthy appearance at the International Exhibition.

Upon entering the Canadian section, on the far wall, one could see the map of New France by Mr. Genest and the large geological map, of Sir William Logan. On the wall to the right, was the table of Canadian Birds and one of the sectional maps of the province of Quebec, drawn by Mr. Jules Taché.

The geological reports of Canada were laid upon a table covered with green velvet with gilt nails ; in the middle was a bronze vase filled with pink wadding which held two splendid specimens of Ottawa phosphate. A second room was also allotted to the province of Quebec which it shared with the Argentine Republic.

In it were hung the cadastral plan of Quebec, by Mr. Paul Cousin ; a map of the Eastern Townships showing the railways ; one of the province of Quebec also showing the railways and showing the minerals of commerce ; one of the projected Laurentian railway one of the Crown Lands Domain, the table of the forest trees of Canada and very fine views of Niagara belonging to Count Viola.

On a second table, similar to the first, were laid the reports of the various Departments of the province of Quebec since 1868 ; those of the Federal Government, the complete collection of our woods and the

Census of 1871. On a third were placed the other books mentioned in the Catalogue. In the first room described above, Count Viola had placed the arms of the Province of Quebec with two English flags on the right and two French flags on the left. In the second department a pearl grey gonfalon, with golden fringe and having above it a red and gold tassel, hung from the ceiling. On it was inscribed the word " Canada ". On the Royal Palace, over the spot in which the Canadian Exhibit was placed, floated a banner bearing, on a blue ground, the arms of the Canadian Confederation.

I give you all these details in order to show you what taste and what tact Count Viola displayed, as representative of the Province.

During my stay in Venice, which was only too short, I had the pleasure of making every effort to give information to the members of the Congress, upon the wealth and resources of Canada and particularly those of the Province of Quebec. Having been invited to give a lecture before the sixth group of the Conference, that of Economical, Commercial and Statistical Geography, I had the honor of seeing in my audience, Mr. de Lesseps, General Thür, Mr. Levasseur of *L'Institut de France*, Mr. de Quatrefages and many other illustrious persons. Colonel Coello, of the Spanish army, occupied the chair.

I submit you this lecture :

CANADA—PROVINCE OF QUEBEC.

Canada, formerly a colony of France, ceded to England by France in 1763, is situated to the north of the United States of which it forms the entire northern frontier for a distance of 1000 leagues.

Formerly divided into Upper and Lower Canada, it has, since 1867, formed a Confederation, under the name of the " Dominion of Canada ".

This Confederation includes the provinces of :

I. Quebec, Old French Canada or Lower Canada.

II. Ontario, Old English Canada or Upper Canada.

III. New Brunswick.

IV. Nova Scotia.

V. Prince Edward's Island.

New Brunswick, Prince Edward's Island, and Nova Scotia—the latter including Cape Breton—are the Maritime Provinces.

VI. British Columbia.

VII. Manitoba and the District of Keewatin.

The two latter are taken from the North West Territories.

The seven provinces and the North West Territories cover an extent greater than that occupied by the United States of North America. Each province, except Keewatin, which has only just been constituted, has its parliament. Moreover they are represented at Ottawa, the capital of the Canadian Confederation, by the House of Commons and the Senate. Each province elects its representatives in the Commons. The executive, that is, the Federal Government, appoints the senators, who sit for life.

. A lieutenant-governor having the power of choosing his advisers, directs its affairs during five years. A Governor General, appointed by England, is at the head of the Confederation.

Immigration has not been so extensive in Canada, as in the United States. Must we admit the reason !

We have no industries. And yet no country in the
world contains greater mineral wealth. Coal, iron
ores—in Canada iron manufactures are protected by a
duty of 25 per cent—copper—there is a protective
duty of 10 per cent—silver, go !, phosphate of lime,
phosphate of aluminium, building stone, marble, asbestos, antimony, lead, sulphur, slate, mercury, mica, natural gas-wells, petroleum, furs, immense forests,
splendid fisheries, inexhaustible water-courses, cereals,
fertile lands, hunting, all exist in profusion. Some
of the ores are found in a purer state than in any other
part of the world. (1)

Separated from France, its mother-country, for
over 119 years, almost the whole population of the
Province of Quebec, one-fourth that of the Lower
Provinces, one-half of Manitoba and that of some parts
of Ontario, speak the French language and are, above
all, anxious to preserve it. Is not this wonderful ?

The decennial census made in 1881 shows that
Canada contains 4,324,810 inhabitants, showing an increase of 680,498 in ten years.

According to our last statistical reports, the population of Canada in 1871 was 3,700,000 that is to say, about
one tenth of the population of France. If, says a Paris
newspaper, the increase in France had been in proportion to that of Canada, it would have gained 7,000,000
inhabitants in ten years. The same newspaper observes
that the excess of births over deaths in France is only

(1) In a scientific essay, published in *La Nature*, Mr. Gaston Tissandier, compares the
natural asbestos found in Canada, with that of Italy which is fibrous and glassy. He says : "It
is the Canadian asbestos, of a fibrous and silky nature, which gives the best results and which
is easily woven and made into felt. The Italian asbestos is difficult to weave, the vitreous
asbestos has no consistency, and crumbles to the touch ; it does not appear capable of being
utilized."

Asbestos is used as cord matting for stopping boxes of steam-engines ; tissues are also
made of it for filtering acids ; boards for the joints of steam engines ; felts for rollers of calend-
ers and for some kinds of electric batteries ; cement for tubing ; paper, &c., &c. Mr. Tissandier
states positively that Mr. Gluk, civil engineer, has discovered an ink with which one may write
on this paper without any fear of the hottest fire destroying the writing.

100,000 per annum. 4,350,933 inhabitants in Canada! This figure, appears to us to be very low considering the area of the Dominion, which is 5,426,014 square *kilomètres.* (1) It is true that Europe contains but 10,000,000 squares *kilomètres.*

This population is confined to certain parts of Canada. In these parts the families are as closely situated as in the most thickly populated districts of France or Italy.

The French language is not only official and side by side with the English language in the Province of Quebec, but it is also official in the Federal Government. It is also used in commercial transactions. The province of Quebec is governed by French laws. Its civil Code is almost altogether based upon the Code Napoléon.

Canada has also its French literature and the French press is composed of about 46 papers.

The chief religion in the Province of Quebec is the Catholic religion, whose priests have considerable authority. By their energy and well appreciated zeal, they have preserved the French language in the country.

Canada is traversed by numerous railways, as is shewn by the maps sent to Venice by the Province of Quebec. One of them, the most important, is the

---

(1) The area of Canada is made up as follows :

Ontario	100,480 square miles
Quebec	193,355 " "
New-Brunswick	27,322 " "
Nova Scotia	21,731 " "
Prince-Edward's Island	2,134 " "
Manitoba	150,000 " "
British Colombia, including Vancouver and other Islands	300,344 " "
North-West Territory	1,863,900 " "
District of Keewatin	300,077 " "
Islands in the Arctic Ocean	31,700 " "
Islands in Hudson's Bay	24,000 " "

North Shore Railway which runs from Quebec to
Ottawa, passing through Three Rivers and Montreal';
another, the Levis and Kennebec is destined soon to
place Quebec in communication with the Atlantic, by
passing through the rich gold bearing regions of Beauce
and the State of Maine.

Only a few weeks ago a powerful syndicate was
formed in England, France and the United States, for
the purpose of building a railway three thousand miles
long. It will unite the Atlantic to the Pacific, have,
some day, its terminus at Quebec, and make the St.
Lawrence the shortest and cheapest route for the car-
rying of grain from the North West and be the great
commercial and strategical artery of the Dominion of
Canada.

One of the greatest and finest rivers of the earth,
the St. Lawrence, flows through Canada. It is navi-
gable for over 300 leagues. Its navigation is acces-
sible to the largest vessels, which are enabled to pene-
trate inland as far as Montreal. This river is 2,413
kilomètres long. Canada is the country the best sup-
plied with rivers and has the largest cataracts in the
world. More than twelve lines of steamships now
enable it to communicate with England, Holland, Belgium, Brazil, the West Indies, &c.

Its climate is essentially healthy. As a distinguished French Canadian author, Mr. Paul de Cazes,
has said : " If every one knows that at St. Petersburg
" the average temperature of the three winter months
" is ten degrees Centigrade, many persons are ignorant
" of the fact that it is never lower than eight degrees at
" Montreal. From the meteorological report of an
" incontrovertible authority, in the latter city the ave-

" rage temperature of January, the coldest month of
" the year, varies between nine and ten degrees, Centi-
" grade. These same reports show that the average for
" July, the warmest month in Canada, varies between
" twenty and twenty two degrees.

" The best proof of the salubrity of the climate is
" the quite exceptional vitality of the population of
" Canada in general, and the truly wonderful increase
" of the French race in particular."

Canada is for North America the country of the
future, for, before long, all the territory of the United
States will be inhabited.

The Washington government is already occupying
itself with immigration ; politicians of that country are
about taking steps to provide for the too great increase
of the population. Canada will then be the only country
in the temperate zone open to immigration. The great
fertility of its soil already causes the population of the
United States to cross the Canadian frontier. Since ten
years, a new province, Manitoba, has been established
in Canada and the district of Keewatin has just been
constituted, both of which have been taken from the
North West territory. The population of the former
is increasing wonderfully.

From the point of view of commercial geography,
the importance of Canada cannot be denied. Before
fifteen years, the prairies of the North West will be the
granary of the world.

Phosphate of lime, which is found in large quantities in Canada, is destined to effect as much good as
that formerly produced by the guano of Peru, and
lands containing these mines are granted on very
favorable terms by the Quebec Government. Our

woods for joiner's work, cabinet-work, ship-building, that used for railway ties, spools, match-boxes, toys and furniture are inexhaustible. With us, bird's eye maple (*acer spicatum*) is used as fuel.

The fisheries of the Gulf and River St. Lawrence are of considerable importance. Whales, porpoises and seals are found there. Cod, mackerel, sardines, trout, salmon, maskinongé, sturgeon, smelts, halibut, bar, eels, shad, pike, pickerel, touradi or white salmon, the most delicate fish, several of which are altogether unknown in Europe, are plentiful in the St. Lawrence or its tributaries.

In some parts of the Gulf, I have seen lobsters sold for two and a half francs a hundred. There are also to be found about fifteen kinds of delicious oysters, which are unknown to European epicures.

To those who would wish to enter into commercial relations with Canada, we offer our wheat, peas, oats, barley and maize, our wool, leather, hemp, copper, wood, mineral waters, phosphates, our meat, fish and game, our live cattle, fish-oils, our horses, which are of excellent breeds, our tan-bark, our agricultural implements, which have been awarded prizes at all International Exhibitions.

To travellers and tourists who love the beautiful, who wish to study nature, who seek health, amusement, and rest, we promise a beautiful, vast and picturesque country, a pleasant life in the midst of an industrious, honest and hospitable population.

This is all that a few minutes chat permit me to tell you of a country which contains in the northern part of America, the vanguard of the Latin race. It is open to all who are fond of labor, integrity, the spirit of enterprise and safe investments. It remains for you, gentlemen, to make this known to your friends and I beg to thank the distinguished personages who surround me, for the marks of their approval which they have just given and for their kind attention.

————

These details, incomplete as they are, for I only had a few hours to prepare, are entered in the proceedings of the Congress of Venice.

I now leave Count Viola to speak. He was kind enough to send me, for transmission to you, the following memorial; which will tell you in warm and heartfelt terms the part played by Canada, and particularly the province of Quebec, at this International Exhibition.

To the Honorable E. T. Paquet,
Provincial Secretary,
Of the Province of Quebec.

Sir,

Having had the honor of representing the Province of Quebec at the International Exhibition of Geography at Venice, I can assure you that, if the Quebec section reached only the second rank as to the number of its exhibits, it certainly was not second in the importance of and the interest taken in its collections, its reports, its geographical and geological maps.

In this report which I have the honor to send you, I will endeavor to show you the results obtained by Quebec at this geographical exhibition. Before commencing, I beg you to be indulgent with respect to the style of this report. I am obliged to write you in a language which is not my own.

" Volgami il lungo studio e il grande amore, "

as said our great poet.

Permit me to briefly recall our preliminary labors.

The Province of Quebec having been officially invited by the Consul General of Italy, residing in Montreal, appointed a local commission to prepare the Canadian Exhibit, consisting of Mr. Eugène Taché and Mr. Faucher de St. Maurice, by order in council bearing date the sixth of May 1881, and on the same day, I had the honor of being appointed your Commissioner at Venice.

Twenty one countries took part in the International Geographical exhibition : France, Austria, Hungary, Belgium, Brazil, Canada (Province of Quebec), Chili, Egypt, Germany, Japan, Greece, England, Italy,

Holland and its Colonies, the Argentine Republic, Russia, Spain, the United States of America, Sweden, Switzerland, Venezuela and the European Commission of the Danube.

I consider it my duty to here state that the comparatively short time, the impossibity, owing to distance, of establishing rapid communication between the local committee in Quebec and your commissioner in Venice, prevented your province from exhibiting as great a variety as it might have done. Nevertheless, I must congratulate my colleagues, Mr. Eugène Taché and Mr. Faucher de Saint-Maurice,for the excellent selection of the articles they sent over.

Both from a scientific and from a commercial point of view, the exhibit of the Province of Quebec gave very important information. Every one here was struck with the perfection of your geological studies, the excellence of your administrative, economical, commercial and statistical works.

Quebec was, moreover, made conspicuous by its historical studies, the progress made in its elementary instruction and by certain books of travel. Of the eight classes composing the exhibition, I can assure you that the Quebec Section honorably distinguished itself in five.

Geology is a science around which many others group themselves. With the assistance of physical meteorological, botanical and geological geography it was destined to hold a high place at the International Exhibition of Venice. By its reports on geological studies, by the works published in Canada from 1842 to 1869, illustrated by the quite recent maps of New-Brunswick, the other geological and geognostical maps and charts, your country took the lead at the Exhibition. I am happy to bear testimony to the superiority of these works, not only for their intrinsic value, but as studies, compared with those exhibited by other countries. The labors of the geological survey of Canada have shown us the importance which your scientific men attach to Geology, Paleontology and Mineralogy. Its explorations have covered an immense extent of territory, and its labors extend from one ocean on the East to another on the West, The profiles which the geological survey of Canada has made of your mining lands, can, by the results obtained, compare favorably with the most complete works of the kind just concluded by the French body of mining engineers and K K Geologichan Reichanstadt of Vienna. We are indebted to your Canadian Geologists for the discovery of several new formations and more especially for the completion of those between the Paleozoics. All the scientific men who were present at the Exhibition, unite in congratulating Canada upon the rapid and unmistakeable progress it has imparted to Geology ; they recognize the energy, tact and perseverance with which your public men have encouraged this science from the beginning and they applaud the intelligent and practical efforts made by your young country, in acquiring a knowledge of its territory.

I beg also to call attention to the remarkable and rapid development of your mineralogical and paleontological collections. The former now consist of 2,479 specimens, divided into twenty eight groups. The latter, which in 1863 comprised 1500 species had in 1879 reached the number of 7,000. We are also indebted to the Geological Survey of Canada for improvements and remarkable changes in the coloring of geological charts. I would especially mention the *di-bianco* reserves. (1)

The last geological Congress, held at Bologna in October 1881, had the honor of adopting your system.

The specimens of phosphate of lime from the Ottawa district were much noticed here in the minereo-logical exhibition. Every one knows how importan: this mineral is in agriculture. We have none at all. I even think it is very rare in Europe, except in Spain. The specimens sent by Quebec, were, with your permission, distributed amongst the Italian museums, schools of agriculture, commercial schools and academies. May the Italian Engineers, who intend to work our mines, carefully study the specimens of phosphate of lime! He who will be fortunate enough to discover some in Italy, will endow his country with the greatest source of wealth.

A very distinguished scientist here who analysed the phosphate from Ottawa county, found in it a large proportion of aluminium. He called it phosphate of aluminium.

The Quebec section may also flatter itself upon have obtained the greatest success with its specimens of woods. It was, I have not the slightest hesitation in saying, absolutely and comparatively the most complete and important ever seen in the Royal Palace of Venice.

Japan and Hungary also exhibited collections of woods. That of Japan was particularly noticeable for its elegance. The Japanese revels in details and with him all labor must overcome some obstacle. Consequently,

_____
(1) Italian words applied to the white lines shewn in the geological colors of the New Brunswick charts.

that section had an artistic appearance about it, but it was not sufficiently complete to give an idea of the forest resources of Japan.

That of Hungary more properly belonged to botanical geography. It was particularly behind the Canadian collection, which contained wood for architecture, ship-building, railways and the commonest as well as the richest of cabinet work.

The varied colors of this collection, the difference in specific gravities were particularly admired.

The button-wood (*platanus occidentalis*), red oak (*quercus rubra*), common poplar (*populus tremuloïdes*), white spruce (*albies alba*), mountain maple (*acer spicatum*), red pine (*pinus resinosa*), bass-wood (*tilia americana*), beech (*fagus ferruginea*), red cedar (*juniperus virginiana*), are all greatly used by us in cabinet-making. These woods come to us from Cadora in Venitia and from Istria. We also get them from Egypt and the East while our cabinet-makers get from America the walnut (*juglans nigra*), maple (*acer saccharinum*), pine (*pinus silvestris*), red oak (*quercus rubra*), &c., &c.

I requested Mr. Faucher de Saint-Maurice to be good enough to send me the most detailed information upon the timber trade of Canada.

I may be mistaken, but I think that in the near future, this trade may be greatly increased with Europe. Our forests are exhausted by the enormous quantities of wood required by railways, either for building them or keeping them in order.

The general use of steam-engines has rendered fuel scarcer. There is a much greater demand for it now than formerly, especially in Italy, where we are obliged to import coal. The vast wealth of the Canadian forests cannot be brought to market as rapidly as it might, owing to want of transport. When the great railways ordered by the House of Commons and the provincial legislature, will be completed, you will be in a position to compete strongly with the other timber supplying countries and you may perhaps secure the monopoly of this trade.

The importation of some kinds, which come in large quantities from Canada, has already attracted the attention of the Italian market ; but in order to be in a position to form a correct opinion upon this important subject, of which Mr. Faucher de Saint-Maurice spoke to me, I have still much information to get, as well as to ascertain, above all, the value of various products in Canadian loading ports, the absolute and comparative value of such and such product, its properties, and the system of weights. I would also have to study the anology between your products and ours.

If such studies would be of any use to your government or to any of the Boards of Trade of your country, I would undertake them with pleasure and I place myself at your disposal.

Maps and plans, I need not say, were the chief feature of the Geographical Exhibition of Venice.

The Province of Quebec was represented by nine.

I. Map of the Province of Quebec showing the Crown Lands.

II. Map of the Province of Quebec, showing the lands conceded up to date and those covered by timber licenses.

III. Map of the Province of Quebec, showing the mineral deposits thrown open to trade.

IV. Map of the Province of Quebec, showing the

railways in operation, under construction and projected.
These first four maps were drawn in the Depart-
ment of Crown Lands under the direction of Mr-
Eugène Taché.

V. Sectional map of the Eastern Townships.
VI. Sectional map of Lake St. John District.
VII. Cadastral plan of the city of Quebec.
VIII. Map of the Lower Laurentian Railway.
IX. Old map of New France by Mr. Genest.

I think, Sir, that some details on this exhibition of
maps, which was very fine, may be of interest to the
government of the Province of Quebec.

The Italian section was anxious to unite everything relating to the history of its cartography. It contained everything, from the planisphere drawn by Fra-
Mauro (1) as well as the charts of the seaports which
preceded or followed it, to the grand military map of
the Topographical Institute of Florence.

Switzerland also gave a history of its cartography,
and so did most of the other countries. France exhibited
the map of its general profile, divided into six plates on a
scale of 1,800,000. The differences in level of the soil
were indicated by the curves of the level, traced
every hundred *mètres*. Switzerland also exhibited a
topographical map on a scale of 1,050,000 ; the level of
the soil was indicated by regular curves every ten
mètres. This spendid work contained about 950 plates.
The Military Topographical Institute of Florence,
which had just completed the geodesical study of the
southern part of Italy, sent us the results, in the shape

(1) Fra-Mauro lived in the fifteenth century. This famous plaalsphore was finished in
1459 at St. Michaels, in the Island of Murano, near Venice.

of a photo-lithographed map on a scale of 1,000,000
and containing 277 plates.

The Geographical and Statistical Institute of Spain
exhibited *La Mapa Topographica de Espagna* on a
scale of 1,050,000. The levels were indicated, at every
twenty *mètres*, by horizontal curved lines and at every
ten *mètres* by intermediate dotted lines.

The Belgian Military Topographical Institute had
maps draw on a scale of 1,040,000, 1,020,000 and
1.000,000. These were marvels of accuracy and were
colored by the photozincographic process.

It must not be forgotten that the majority of these
very detailed studies only relate to regions of limited
extent and that these different works, conducted very
slowly, are the result of the united efforts of a great
many scientific institutions. It is no wonder, therefore,
if Canada, confining its efforts to acquiring in a short
time, a knowledge of its immense territory did not
produce at the exhibition topographical works, as mi-
nute in their details as those of other countries. Nevertheless, the jury was of opinion that the Province of
Quebec has extensively carried out its topographical
work, in the interest of economy, agriculture, commerce
and statistics.

The nine maps, as well as the reports of the
public Departments which it sent to the Venice exhibition, amply proved this to the satisfaction of all.

It is not within the limits of a report such as this, that
I can give the results of the careful examination we
made of the Departmental Reports of the Province of
Quebec and the statistical works of the Government of
the Dominion.

We deeply regretted that we were unable to have, at

our exhibition, the results of your last census, but we could admire that of 1871 and it is beyond a doubt that the statistics it contains are worthy of the highest praise.

Canada may well be proud of its Statistical Bureaus ; it may also be proud of the increase of its population.

Such an increase is always an evidence of the prosperity of the country, so that the results given by the last census are a happy omen.

The population of Canada which, in 1871 was composed of 3,718,747 inhabitants, is now 4,350,933 ; which makes an increase in ten years of 632,188 inhabitants or 17 per cent.

A very valuable work, bearing the modest title of *Notes sur le Canada*, was much admired at the Exhibition. " Read M. Paul de Cazes book," said a friend to me quite recently, "and you will become perfectly acquainted with the Canadian Confederation."

In fact, M. de Cazes' work gives very valuable information upon your country. It is a safe guide for those who wish to st idy Canada or enter into commercial relations with it and I think it would be a wise policy to make it better known abroad.

We were all struck with the clear, ample information contained of the Departmental Reports of the Province of Quebec and of the Dominion. Those which relate to Education and Agriculture show the progress they have made; and we, at once, made ourselves acquainted with the cultivation of new products, the ever increasing settlement of your lands, and the working of your forests, mines and fisheries. Your Customs' tables show that the exports and imports are becoming equalized and the maritime movement in your ports places your marine amongst the first ranks.

In one of these reports, I see that the production of beet root sugar and the cultivation of the vine in Canada are spoken of. I need not tell you that, with us, beet sugar has almost entirely superseded colonial sugar. To day the former product rules the European Market.

France, Germany, some parts of the Austrian Confederacy, etc., manufacture it in large quantities, while colonial sugar, which was once a monopoly of Holland, is only used for special purposes. I am happy to notice that Canadian farmers are taking an interest in a product, which has not yet become general in the New World. This industry is destined to give employment to a considerable number of persons and realize large profits.

I am also delighted to see that the vine is being cultivated in Canada. The satisfactory results which it has given, prove that your soil and climate are favorable to its growth and the Government cannot give too much encouragement to this patriotic work. The wines of the United States are beginning to be known in Europe and why should not Canadian wines become known also? The phylloxera and various other diseases of the vine decimate our vine-yards in Europe. Even here in Italy, we have the *pellagra*, a direful disease which ruins our agricultural classes and increases the number of the inmates in our penitentiaries and lunatic asylums. The *pellagra* has spread to a terrible extent since the day when the *crittogama* ravaged our vine-yards and a considerable portion of our rural population has been compelled to give up using wine. Insufficiency of food which is, with us, the chief cause of the *pellagra*, found a powerful auxiliary in the use of a diseased wine which no longer assisted digestion and impoverished the blood of persons whose strength was already exhausted by labor.

As one of our Italian songs says :

L'aqua fa male
Il vino fa cantare.

Your agricultural, your forest and mineral wealth, your fisheries, your new industries, can but be developped and become known by increased facilities for transport and communication.

The jury notice, with pleasure, the great extent of the railway network which covers the territory of Canada. Rapid locomotion is, without doubt, the most important factor of production and commerce. Canada seems convinced of this; and the International Geographical Congress of Venice notices with pleasure that your country has commenced the building of a railway which, starting from British Columbia, will cross the vast territories of the West, pass by Montreal and Quebec, where it will join the Intercolonial, unite the Pacific with the Atlantic and thus place old Europe in direct communication with the far East. The Canadian Pacific Railway, joined to the Intercolonial by the North Shore road, will, we are convinced, rank amongst the most celebrated undertakings of the 19th century.

The teaching of Geography occupied a prominent position at the Venice Exhibition. It was represented by works, instruments, maps, collections, &c. You have no idea of the progress made in elementary Geography since some years. Formerly Geography was only taught mechanically in the Colleges. It was studied more as a curiosity than as an absolute necessity. Now it is very different. The child's work is rendered easier by means of interesting books, of maps in relief and ingenious mechanisms. From what we were enabled to see and observe at the Exhibition. Russia is the country which has made the most remarkable progress in elementary geography. Its pedagogical museum was the most important of all those in the Royal Palace of Venice. It contained a number of elementary apparatus, each more curious than the other. geographical toys for children, ethnographical albums. &c., &c.

The advantage of this system of instruction was shown by geographical maps, admirably executed by the pupils. In this section were also to be found geographical machines intended to explain to the children the movement of the stars, the planetary systems, the rotatory movement of the earth and its movement around its orbit.

Hungary exhibited chromo-lithographic plans, intended for attracting the attention of the children to celestial and terrestrial phenomena. To understand this thoroughly it was sufficient to follow attentively the remarks made by the professor. He described the plates, explained the phenomena which they represented thus impressed the lesson much more easily on the minds of his pupils.

Many illustrated books and narratives of travels formed part of the seventh and eight classes. In the Quebec Section a work which was particularly admired was " The Geography and History of the British Colonies," a splendid publication, illustrated with 72 engravings and published in Montreal by Mr. Lovell. The works of Mr. Faucher de Saint-Maurice appeared to the jury to be very remarkable and I consider them very useful for seamen and students who wish to learn about the Gulf of St. Lawrence, its Islands and the Maritime Provinces of the Canadian Confederation.

The author is one of those who form that brave group of travellers who have come from all parts to help us in adding to the wealth of the International Exhibition of Venice. Each one was careful to being his notes, and relations of facts: Most of them had added articles gleaned and collected in far distant and almost unknown regions, maps of newly explored territories made according to fairly correct observations, photographic views of native habitations, savage huts, charming landscapes, drawings from nature and collections of the fauna and flora of the country visited. Thus we all had an opportunity of admiring the collection of the hardy Arctic explorer Mr. Nordenskjold ; it was exhibited by Sweden and was made up in a very intelligent manner. This traveller had omitted nothing which could give one an idea of the social and material life of the Esquimaux and other races visited by the *Vega* on her polar voyage. By looking at this collection, one could at once get an idea of the habits, occupations, and degree of civilization of these mysterious inhabitants of the ice kingdom.

The Egyptian exhibit was no less interesting than that of Sweden. One remained struck with amazement before the glass cases, containing swords, helmets, buck-, lers, kangiars, knives and rich dresses of the Dar Fur.

Arms, instruments, furniture, stuffs, ornaments, coin, jewels, objects of art, idols of the savage tribes of Bahr-el-Gazal, of Bahr-el-Gebel, Djour, Bongo, Niam-Niam, Tiki-Tiki were spread out in profusion, while beside them were seen the commercial products of these countries, such as ivory, ostrich feathers, tiger, hippopotamus and crocodile skins, products which are so sought after in Europe and which these tribes barter for pipes, glass-beads, gun-powder and brandy.

I have just gone rapidly over the classes of the Venice Exhibition in which the Province of Quebec appeared and I repeat, that notwithstanding the short space of time in which it had to prepare, it made a very creditable appearance. Nevertheless, I could not help regretting, in Venice, the absence of certain collections which I had so much admired during my stay in Canada : The ethnographical museum of Laval University formed by Doctor Jean Charles Taché, who has a great reputation ; the botanical collection of the same University ; certain series of the geological museum in Montreal, the beautiful specimens of geology, ichthyology and ornithology which belong to Laval, to the Quebec Litterary and Historical Society and to Mr. J. M. Lemoine of Spencer Grange would have made a worthy appearance at this International exhibition.

By its presence here, by the success it has obtained Quebec, like the other countries which participated in the exhibition, proves the importance which should be attached to geography. Are not the sciences its tributaries ? and to-day, more than ever, man studies with ardor, that great book which we call the World.

My task is now performed, but before offering my respectful homage, allow me to communicate the decisions of the International jury, respecting the share which the Province of Quebec has taken in the Venice exhibition.

The following prizes were placed at the disposal of the jury for distribution amongst exhibitors.

48 First class medals with diplomas ;
96 Second class medals with diplomas ;

144 diplomas of honor.

At the last general meeting of the International Geographical Congress of Venice, Doctor George Schweinfurth, read out, in solemn conclave, the names of the fortunate exhibitors. The Canadian Section received the following awards.

CLASS III.

GEOLOGICAL GEOGRAPHY.

Letter of distinction to the Geological Commission of Canada.

CLASS IV.

ECONOMICAL, COMMERCIAL AND STATISTICAL GEOGRAPHY.

First class Diploma of honor to the Government of the Province of Quebec.

CLASS VII.

METHODOLOGY, TEACHING AND DIFFUSION OF GEOGRAPHY.

Honorable mention to the Government of the Province of Quebec.

CLASS VIII.

GEOGRAPHICAL EXPLORATIONS AND TRAVELS.

Honorable mention to Mr. Faucher de Saint-Maurice.

The official and complete list of the prizes, awarded by the International jury of the Venice Exhibition, was published in No. 11 of the *Diario* which I have the honor to send you with this report.

As to the sending of the prizes, the Secretary General of the Congress, Mr. Dala Vedova, writes me, from Rome, that the diplomas are at Turin, awaiting the signature of His Royal Highness the Duke of Genoa. I will have the honor of forwarding them to you during the course of next May.

My work is now concluded and I would be happy if my feeble efforts had contributed to establish scientific and commercial relations between your country and mine. Canada is beginning to be known in the scientific circles and business establishments of Italy. The exhibition just held at Venice has a great deal to do with this. I will be rewarded beyond all my expectations if my modest report has the effect of drawing closer together the commercial interests of my country and that beloved Canada which has left such agreable impressions on my mind, from my too short stay amongst the Canadians.

I particularly visited the city and Province of Quebec where I received the warmest of welcomes. In return, permit me to assure you that I will do all I can in Italy to make your government, your institutions, your products better known and to increase the prosperity of friends of whom I retain the pleasantest of recollections.

Count JEAN-BAPTISTE VIOLA,

Commissioner of the Province of Quebec
at the International Geographical Exhibition of Venice.

Venice, 3rd November 1881.

I take the liberty of adding a few words to the very flattering remarks contained in the report of His Excellency, Count Viola. In this study, which is full of sympathy for us, he points out the important part taken by Quebec at the International Exhibition of Venice. Our public instruction, our teaching of geography, the reports of our various departments, our geological labors attracted the attention of those who took part in this scientific Congress. Canada and Quebec may well be proud of the prizes which they gained.

A paper published under the name of *Venezia e il Congresso*, of which only one number was published, contained an excellent article on our province. This short study was from the graceful pen of your Commissioner Count Viola.

What can I say of the boundless hospitality of the city of the Doges ?

During fifteen days, Venice, in holiday attire, with its syndic Count Allighieri Dante at its head, welcomed its guests from all quarters of the word. Balls, dinners, regattas and senerades in gondolas, a royal gala night at the *Fenice*, an official visit to the University of Padua, a night concert on the Place of St. Mark, which was illuminated by electricity, fire-works, conferences, visits to churches, palaces, museums, arsenals, manufactures of bronzes, mirrors, glass-work, horticultural exhibitions, in fact everything that a rich, clever, artistic, amiable and hospitable city could think of for pleasing its guests, was placed at the disposal of the members of the Congress.

The leading merchants and bankers of Venice are disposed to open business relations with the Dominion of Canada and particularly with the Province of Quebec.

I had the honor of holding lengthy conversations with some of them on the subject. The members of the Board of Commerce and Arts of Venice did me the honor of sending me interesting works on the statistics of the navigation and commerce of the Adriatic. I have sent them to the Library of the Legislature, Quebec. The Board of Commerce of Venice, is desirous of opening communications with the Board of Trade in Canada, and one of its members, Mr. Councillor Eugenio Vio, kindly offered his services to give us all the information we may require, with respect to the Italian Trade. Several merchants also expressed a wish to have samples of our leather. If our hides, undressed leather and sole leather suit, they may be the object of a profitable trade between Quebec and Italy. The same applies to our tan-bark extracts which would find a ready sale there. White petroleum well refined, our prepared phosphates would also find a ready and profitable market. Many questions were asked as to the quality of Canadian coal, and its advantages as compared with other coal. Almost all the coal used in Italy comes from Styria. It is worth about 80 per cent of the value of Glasgow coal.

They are also anxious to know the prices, on board, of our pork, wheat, barley and especially our oats. For a return cargo we would have all the products of Italy such as fruits, oils and wines, amongst which I may mention the celebrated Braganza vintage, which is as good and as dry as the best Xerès. The owner of this celebrated vineyard told me that he could sell this wine, bottled, packed and placed on board at Genoa for 20 francs per dozen and 24 francs for the best.

Quebec has everything to gain by making itself known abroad ; we have everything needed for success ; everything which makes a country rich and prosperous. During the International Congress, our province, while showing itself deserving of prizes which more than one country envied it, was praised for its exhibit by the London *Times*, as well as by the leading papers of Italy, Austrian, Switzerland, Denmark, Sweden, Norway, Russia and France. Among the latter I may mention *Le Temps*, in which M. Levasseur a member of the Institute and one of the most illustrious fellows of the Geographical Society of Paris, published an interesting series of articles on the Venice Exhibition.

Before concluding, permit me to call your attention to the services rendered to the province, in connection with this exhibition, by M. Eugène Taché, assistant-Commissioner of Crown Lands, and Canadian Commissioner for the Venice Exhibition ; Messrs. Ferdinand Borsari, Alphonse Audinot, Doctor Count Braganza, all members of the International jury for the Canadian Section ; the members of the Board of Commerce of Venice : MM. Eugenio Vio and Ferrari. If the Quebec exhibit met with such success at Venice, it is due to their tact and to the judgment of the members of the jury. I need not tell you of the zeal displayed by His Excellency Count Viola ; the interests of Quebec could not be in better hands. Our representative at Venice resided for some time in our capital. He knows our country thoroughly and like all who have come here, like the officers of the war-ships who honor us with their visits, the members of the French Commission to Yorktown whom we have had the pleasure of seing recently, Count Viola has shown that he has not forgotten us. The memorial he has written you is a proof of it.

Every member of the Congress was allowed to speak his own language, but I am happy to be able to tell you that the real language of the International Geographical Exhibition of Venice, was French.

I remain, with the greatest respect, your obedient servant ;

FAUCHER DE SANT-MAURICE

Commissioner for the Province of Quebec,
at the International Geographical Exhibition of Venice.